Secret Worlds

Too Fast To See

Kim Taylor

Delacorte Press

Published by Delacorte Press
Bantam Doubleday Dell Publishing Group, Inc.
666 Fifth Avenue, New York, New York 10103
This edition was first published in Great Britain
in 1989 by Belitha Press Limited.

Consultant: S. Patricia Manning

Manufactured in Italy for Imago Publishing
October 1991
10 9 8 7 6 5 4 3 2 1

Library of Congress Cataloging in Publication
Data

Taylor, Kim.
Too fast to see/Kim Taylor.
p. cm.—(Secret worlds)
ISBN 0-385-30218-5
ISBN 0-385-30219-3 (lib. bdg.)
1. Motion perception (Vision)—Juvenile
literature. 2. Speed—Juvenile literature.
3. Animal locomotion—Juvenile literature.
I. Title. II. Series: Taylor, Kim. Secret worlds.
BF245.T38 1991
591.1′852—dc20 90-3334
CIP
AC

SOME THINGS IN NATURE HAPPEN SLOWLY, WHILE others are too rapid for your eyes to follow. When a big bird glides by, you can see how it holds its wings, but the wings of a fly, or even of a little bird, move so quickly that all you see is a blur. If you want to know what is happening when things move too fast to see, you have to look at special high-speed photographs. On the following pages there are pictures of things you cannot see properly with just your eyes. You need some help. A camera can stop time for you. So the next time you see a fly fly or a drop drop, remember what you have seen here and you will know what is *really* happening.

Buzzing beetles

A HEAVY, CHUNKY BEETLE DOESN'T LOOK AS IF IT *CAN* fly. Where are its wings? They are hidden beneath the hard wing cases, called elytra, on its back. When a European cockchafer beetle wants to fly, it first spreads out its antennae like fans (*below*). Then it opens its elytra so that its wings are free. The wings move very swiftly—about thirty times a second. They are like transparent paper or plastic with thin struts to stiffen them.

Can you see the struts in the wings of this flying beetle (*opposite*)? If there were no struts, the wings would crumple and the beetle would never get off the ground. But here it is, buzzing happily near the top of a tree. When it lands, the wings are carefully folded under its elytra.

Flapping flies

THESE DAMSELFLIES (*OPPOSITE*) LIVE NEAR A STREAM. The damsel in the air has its four wings spread and is hovering like a helicopter before landing. Of course, its wings do not spin around like a helicopter's rotor, but they do flap up and down too fast for your eyes to follow.

What is happening in the picture below? You might think that it is a string of eight delicate flies hanging in the air, but it isn't. It is eight shots of one beautiful lacewing fly taking off from a bud. Can you see that the wings change position in each part of the picture as the lacewing flies?

Taking off and landing

THE PICTURE OPPOSITE SEEMS TO SHOW THREE European robins taking off from a post. But does it? Look carefully, and you will see that the beak of one of the birds shows through the tail of the one in front. Something funny is going on here. In fact, the picture is made up of three separate high-speed shots of one bird taking off in a hurry. See how its wing feathers are spread and how it pulls up its feet.

Most airplanes pull their wheels up after takeoff. They do this because wheels would slow them down in the air. Birds are just the same. Feet get in the way when they are flying. At landing time, down come the wheels and down come the feet. The little owl, below, is getting its feet ready for landing. Its wing feathers are also spread wide, just like a landing airplane spreading its wing flaps.

Bathtime

WHEN A BIRD WANTS A BATH, IT HAS TO FIND WATER that is not too deep. The thrush (*above*) is standing in water up to its breast and shaking its head from side to side. Water is being thrown up into the air by the bird's beak.

Can you see something odd about the thrush's eye? It has a sort of milky look. This is because it is covered by a special thin eyelid. Normally this eyelid moves so quickly that you can hardly see it.

After throwing water in the air with its beak, a bathing bird like this little wren (*opposite*) flutters its wings sending spray everywhere.

Plop goes a drop . . .

THE NEXT TIME YOU ARE STANDING BY A POND OR puddle and it starts to rain, don't run for cover. Stay and watch what happens. As each raindrop hits the pond, a little spurt of water comes up (*above left*). Sometimes you hardly notice the drops coming down. All you see are the spurts coming up. It looks as if the water has become spiky (*above right*).

If it is really heavy rain (and you have an umbrella),

stay and watch longer because you will see something different happen. As each large drop plops into the pond, instead of a spike, a bubble is formed. Usually the bubble lasts no time at all, but sometimes it drifts for a second or two before it bursts. These pictures show how a drop makes a bubble. As the drop plops, it sends up a ring of fine spray. This turns into a ring of water shaped rather like a crown. The top of the crown then collapses inward until, presto! you have a bubble with a few drops on top.

. . . and splash goes a frog

A FROG SITS HIDDEN BENEATH SOME FLOWERS (*BELOW left*). It has been there for quite a while. Suddenly, and without warning, it leaps across the grass (*below right*). See how it partly closes its eyes as it jumps and how little drops of water from the wet grass follow it. It is heading for the pond; you know how frogs like water. As it dives headfirst into the pond (*opposite*), water is thrown up in all directions and you can just see the frog's eye, still partly closed to protect it.

Sticky tongues

HERE IS A FROG LEANING FORWARD TO EAT A BEETLE larva (*above*). You might expect it to open its mouth and pick up the larva, but it does not. Instead, out shoots a pink tongue that flaps down onto the frog's dinner (*below*). In a flash, tongue and dinner disappear into the frog's mouth, leaving you wondering whether you really saw anything at all.

A toad has a longer tongue than a frog, so it does not have to lean forward to get its dinner. Instead, it just sits there, looking as if it was not interested in anything. If an insect moves into range, the tongue flips out with a little pop (*above*) and the insect is gone. This is a very clever vanishing trick, which works only because the end of the toad's tongue is sticky. If you look very carefully, you might just see the insect stuck to the tongue at the back of the toad's mouth (*below*).

What a tongue!

YOU'VE JUST SEEN HOW USEFUL A TONGUE CAN BE. The toad's tongue was longer than the frog's, but even so, toads and frogs find it difficult to catch flies. This is because toads and frogs have to get really close to the flies and would frighten them away.

Chameleons do not have such a problem because their tongues are enormously long. Here (*opposite*) you can see a chameleon "shooting" a fly from nearly six inches away. The end of its tongue is sticky, so the fly cannot escape. It is hauled back into the chameleon's mouth as a tasty morsel. How does the chameleon's long tongue fit into its mouth? It isn't coiled up like its tail but is neatly folded instead—rather like an accordion.

The next time you see a fly sitting on a wall, try putting your hand close to it. If you move your hand quickly, the fly will buzz away at once. Now try moving your hand very, very slowly toward the fly. Can you get as close as a chameleon would need to be to catch the fly? Now try getting as close as a *toad* would need to be. That's *much* more difficult.

Running

HERE IS A DOG RUNNING WITH A STICK IN ITS MOUTH (*left*). The dog is moving so fast that the ground and the trees and the sky behind it are just a blur. The next time you see a dog running fast, try to watch what it does with its feet.

Ponies can walk, trot, canter, and gallop. Here (*above*) is a foal galloping. You can see all four of its feet close together as it sails through the air. The foal's mother is trotting. She has two feet on the ground and two off the ground.

Are the feet that are off the ground on the same side or on opposite sides of the horse's body?

Jumping

LOOK AT THE CAT JUMPING, FIRST UP AND THEN DOWN (*opposite*). Before it starts to go up, see how its hind legs are bent. Then they straighten and the cat goes flying upward. Spiders can jump, too. The one above is leaping from a twig to a fir cone. The picture shows three stages of the jump. Can you see that the spider has spun itself a lifeline? This will hold the spider if it does not land safely. The spider can then climb up the line and try to jump again.

Buzz off

CAN YOU GUESS WHAT THIS IS? IT HAS SIX BLACK LEGS, a pair of papery wings, and two shiny red wing cases with black spots on them. Red with black spots: That's the clue. Could it be a ladybug? Yes, that is what it is. Flying along like a miniature helicopter, it comes zooming out of the page toward you. Look out. Duck, and it might just pass over your head.